UAV PILOT
LOGBOOK
2ND EDITION

HOLDER'S NAME:

_____

Logbook number _____

Entries from _____

Through _____

HOLDER

Holder's Operator Certificate	Holder's Address
Certificate	
Number	
Date / place of initial qualification	
(Space for address change)	*(Space for address change)*

CONTENTS

Notes on the use of this book:

Use Part 1 to record aircraft and equipment characteristics of any aircraft flown.

Use Part 2 to record all flights.

Use Part 3 to record flight and ground training courses taken and signature-certification of training programs completed.

PART 1: AIRCRAFT AND EQUIPMENT DATA

AIRCRAFT IDENTIFICATION DATA

Identity data of aircraft recorded in this logbook:

1. Make _____ Model _____ Year of Manufacture _____

Registration Number _____ Type (e.g. Fixed-wing, Rotary, LTA) _____ Date Acquired _____

Owner Name / Address _____

2. Make _____ Model _____ Year of Manufacture _____

Registration Number _____ Type (e.g. Fixed-wing, Rotary, LTA) _____ Date Acquired _____

Owner Name / Address _____

3. Make _____ Model _____ Year of Manufacture _____

Registration Number _____ Type (e.g. Fixed-wing, Rotary, LTA) _____ Date Acquired _____

Owner Name / Address _____

AIRCRAFT IDENTIFICATION DATA, CONTINUED

Identity data of aircraft recorded in this logbook:

4. Make _____ Model _____ Year of Manufacture _____

Registration Number _____ Type (e.g. Fixed-wing, Rotary, LTA) _____ Date Acquired _____

Owner Name / Address _____

5. Make _____ Model _____ Year of Manufacture _____

Registration Number _____ Type (e.g. Fixed-wing, Rotary, LTA) _____ Date Acquired _____

Owner Name / Address _____

6. Make _____ Model _____ Year of Manufacture _____

Registration Number _____ Type (e.g. Fixed-wing, Rotary, LTA) _____ Date Acquired _____

Owner Name / Address _____

AIRCRAFT IDENTIFICATION DATA, CONTINUED

Identity data of aircraft recorded in this logbook:

7. Make   Model   Year of Manufacture

Registration Number   Type (e.g. Fixed-wing, Rotary, LTA)   Date Acquired

Owner Name / Address

8. Make   Model   Year of Manufacture

Registration Number   Type (e.g. Fixed-wing, Rotary, LTA)   Date Acquired

Owner Name / Address

9. Make   Model   Year of Manufacture

Registration Number   Type (e.g. Fixed-wing, Rotary, LTA)   Date Acquired

Owner Name / Address

AIRCRAFT IDENTIFICATION DATA, CONTINUED

Identity data of aircraft recorded in this logbook:

10. Make   Model   Year of Manufacture

Registration Number   Type (e.g. Fixed-wing, Rotary, LTA)   Date Acquired

Owner Name / Address

11. Make   Model   Year of Manufacture

Registration Number   Type (e.g. Fixed-wing, Rotary, LTA)   Date Acquired

Owner Name / Address

12. Make   Model   Year of Manufacture

Registration Number   Type (e.g. Fixed-wing, Rotary, LTA)   Date Acquired

Owner Name / Address

POWER SOURCE DATA

("Out," with checkbox, designates a device no longer in use)

Batteries for Aircraft (name / number) _____

	Out: ☐	Out: ☐	Out: ☐	Out: ☐
Battery Designation				
Make / Type				
Date Acquired				
mAh / Vmax				

Batteries for Aircraft (name / number) _____

	Out: ☐	Out: ☐	Out: ☐	Out: ☐
Battery Designation				
Make / Type				
Date Acquired				
mAh / Vmax				

Batteries for Aircraft (name / number) _____

	Out: ☐	Out: ☐	Out: ☐	Out: ☐
Battery Designation				
Make / Type				
Date Acquired				
mAh / Vmax				

Batteries for Aircraft (name / number) _____

	Out: ☐	Out: ☐	Out: ☐	Out: ☐
Battery Designation				
Make / Type				
Date Acquired				
mAh / Vmax				

Batteries for Controller (name / number) _____

1. _____ Out: ☐
2. _____ Out: ☐
3. _____ Out: ☐

Batteries for Controller (name / number) _____

4. _____ Out: ☐
5. _____ Out: ☐
6. _____ Out: ☐

Liquid-fuel for Aircraft (name / number) _____

Type: Full Tank Quantity:

Buoyancy Gas for LTA Aircraft (name / number) _____

Type: Full Aircraft Quantity:

POWER SOURCE DATA, CONTINUED
("Out," with checkbox, designates a device no longer in use)

Batteries for Aircraft (name / number) _____

Battery Designation	Out:☐	Out:☐	Out:☐	Out:☐
Make / Type				
Date Acquired				
mAh / Vmax				

Batteries for Aircraft (name / number) _____

Battery Designation	Out:☐	Out:☐	Out:☐	Out:☐
Make / Type				
Date Acquired				
mAh / Vmax				

Batteries for Aircraft (name / number) _____

Battery Designation	Out:☐	Out:☐	Out:☐	Out:☐
Make / Type				
Date Acquired				
mAh / Vmax				

Batteries for Aircraft (name / number) _____

Battery Designation	Out:☐	Out:☐	Out:☐	Out:☐
Make / Type				
Date Acquired				
mAh / Vmax				

Batteries for Controller (name / number) _____

1.	Out:☐
2.	Out:☐
3.	Out:☐

Batteries for Controller (name / number) _____

4.	Out:☐
5.	Out:☐
6.	Out:☐

Liquid-fuel for Aircraft (name / number) _____

Type: _____ Full Tank Quantity: _____

Buoyancy Gas for LTA Aircraft (name / number) _____

Type: _____ Full Aircraft Quantity: _____

POWER SOURCE DATA, CONTINUED

("Out," with checkbox, designates a device no longer in use)

Batteries for Aircraft (name / number) _____

Battery Designation	Out:	Out:	Out:	Out:
Make / Type				
Date Acquired				
mAh / Vmax				

Batteries for Aircraft (name / number) _____

Battery Designation	Out:	Out:	Out:	Out:
Make / Type				
Date Acquired				
mAh / Vmax				

Batteries for Aircraft (name / number) _____

Battery Designation	Out:	Out:	Out:	Out:
Make / Type				
Date Acquired				
mAh / Vmax				

Batteries for Aircraft (name / number) _____

Battery Designation	Out:	Out:	Out:	Out:
Make / Type				
Date Acquired				
mAh / Vmax				

Batteries for Controller (name / number) _____

1. _____ Out:
2. _____ Out:
3. _____ Out:

Batteries for Controller (name / number) _____

4. _____ Out:
5. _____ Out:
6. _____ Out:

Liquid-fuel for Aircraft (name / number) _____

Type: _____ Full Tank Quantity: _____

Buoyancy Gas for LTA Aircraft (name / number) _____

Type: _____ Full Aircraft Quantity: _____

EQUIPMENT

Make / Model / Identification Number / Purpose of equipment recorded in this book (e.g. cameras, lenses, sensors) (The term "Out," with check-box, designates a device no longer in use)	Date Acquired
1.	Out: ☐
2.	Out: ☐
3.	Out: ☐
4.	Out: ☐
5.	Out: ☐
6.	Out: ☐
7.	Out: ☐
8.	Out: ☐
9.	Out: ☐
10.	Out: ☐

EQUIPMENT, CONTINUED

Make / Model / Identification Number / Purpose of equipment recorded in this book (e.g. cameras, lenses, sensors) (The term "Out," with check-box, designates a device no longer in use)	Date Acquired
11.	Out:☐
12.	Out:☐
13.	Out:☐
14.	Out:☐
15.	Out:☐
16.	Out:☐
17.	Out:☐
18.	Out:☐
19.	Out:☐
20.	Out:☐

PART 2: PILOT FLIGHT LOG

Notes on the recording of pilot flight data (on even-numbered pages) in this logbook:

<u>By columns across the top of each page / per flight:</u>

Flt. No. Enter a sequential number for each flight (e.g. 1, 2, 3, or F1, F2, F3, etc.).

Date Enter the current year in the space provided at the upper left-hand corner of each flight data page. Enter the day and month on which each flight commences in the spaces given.

Aircraft Enter the make, model and registration number of the aircraft flown.

Flight Location Enter the takeoff location and time and the landing location and time. Note that the takeoff and landing may occur at the same place and need be entered only once for both phases of flight. Remark whether you use local time or UTC.

Flight Duration Enter the duration of each flight, in hours and minutes or hours and tenths, in the column that corresponds to the type of aircraft being flown and again in the column marked "Total." In this manner it is possible to record separately the pilot's experience with Fixed-Wing, Rotary and Lighter-Than-Air (LTA) UAV aircraft and to sum the total flight time.

Function Enter the duration of the flight, in hours and minutes or hours and tenths, in the column that corresponds to the role performed by pilot during the flight. In this manner it is possible to record whether the flying was done as Operator in Charge (OIC) / Solo, or while receiving Training or acting as Instructor, separately from the total flight time.

Data Enter the number of landings performed during the flight in this section, as well as the distance flown. Only landings will be totaled at the bottom of the page.

Notes on the recording of pilot flight data (on even-numbered pages) in this logbook, continued:

Totals across the bottom of each page:

Totals this page: Sum the durations and landings entries from the columns above and enter these data in the spaces provided.

Totals brought forward: Enter the '**Totals to date**' data from the *previous page* in the spaces provided.

Totals to Date   Sum the 'Totals this page' and the 'Totals brought forward' entries and enter these data into the spaces provided. Note that the resultant '**Totals to date**' for Fixed Wing plus Rotary plus LTA flight time should equal the entry for 'Total' time (bold-bordered box). The '**Totals to date**' for 'OIC/Solo' plus 'Training Received' and 'Instructor' flight time should equal the entry for 'Total' time as well.

Sign the page in the lower-left corner in the space provided to attest to the fact that the entries are true and accurate.

An example is provided on the next page.

FLT. NO.	DATE Yr 2017 Dd / Mm	AIRCRAFT Make/Model Registration No.	FLIGHT LOCATION Takeoff Time *L*	FLIGHT LOCATION Landing Time *L*	FLIGHT DURATION Fixed Wing	FLIGHT DURATION Rotary	FLIGHT DURATION LTA	FLIGHT DURATION Total	FUNCTION OIC/Solo	FUNCTION Training Received	FUNCTION Instructor	DATA Ld.	DATA Dist. *meters*
1	01/03	*DJI P3P* N12345	*Family farm, Anywhere, CO* 0800	0830		0:30		0:30	0:30			1	600
2	02/03	*My Own* N54321	*364609N 1041317W* 1210	1245	0:35			0:35			0:35	1	550

Example page:

In the example above the first two entries of a page and separately, below them, the corresponding totals, are shown as if the entire page had been filled-out and eight flights flown and entered.

The pilot has flown both Fixed-wing and Rotary aircraft and has logged the times separately. The duration of each flight has also been entered into the 'Total' column, so that the combined time of all aircraft flown can be calculated.

This pilot has entered the 'Totals this page' in the spaces provided at the bottom and has copied the corresponding 'Totals to date' from the previous page and entered them here as 'Totals brought forward.'

The Fixed-wing time flown to date is therefore shown to be 7 hours 45 minutes, the total Rotary time to date is 5 hours 55 minutes and the pilot's absolute total time is 13 hours 40 minutes for all aircraft together.

Note that this pilot has flown 1 hour 55 minutes while giving instruction and 11 hours 45 minutes as OIC / Solo, which also sum to 13 hours 40 minutes. He has annotated "L" (local time), "meters" at the top as his units.

	Fixed Wing	Rotary	LTA	Total	OIC/Solo	Training Received	Instructor	Ld.	Dist.
I certify that the entries in this log are true. Totals this page:	4:15	3:40		7:55	6:00		1:55	8	
John Smith Totals brought forward:	3:30	2:15		5:45	5:45		0	8	
Totals to date:	7:45	5:55		13:40	11:45		1:55	16	

Flt. No.	Batteries / Fuel	P-f. Ins. C. Cal.	Technical Problems Found / Corrective Action, Status	Remarks and Endorsements
1	*Batt. A, volts*	*Y*	*Aircraft would not fly a straight line.*	*Satisfactory video data obtained for client after*
	17.1 – 14.6	*Y*	*Made IMU and controller stick calibrations.*	*performing calibrations.*

Example and Notes on the recording of technical data and remarks (on odd-numbered pages) in this logbook:

Flt. No.   Enter the corresponding flight number from the previous page. The entries on this page support the flights entered on the previous page (or page "above"). The flight number entered here, therefore, refers directly to a flight recorded above; in this example: flight number 1.

Batteries / Fuel   Enter amount of fuel uplifted and burned (for liquid-fueled aircraft) or battery charge and discharge values (in volts or % charged). Be sure to state the battery designation (e.g. name, number) recorded in the Power Source Data pages. In this example the pilot recorded that battery "A" had a charge of 17.1 volts prior to flight and a post-flight charge of 14.6 volts remaining.

P-f. I. / C. Cal.   Pre-flight Inspection / Compass Calibration: remark, using a crew-member's initials or a "Y" for Yes, that a pre-flight inspection was accomplished. Remark also (Y / N) whether a compass calibration was performed or not. In the example here both remarks are "Y."

Technical Problems Found / Corrective Action, Status
     Enter a brief description of any mechanical defect or incident that occurred during the numbered flights (or pre- flight). Enter also a description of any repair or fix accomplished to resolve the problems. Comment about any test flight requirement.

Remarks / Endorsements
     Enter pertinent observations about the flight (e.g. observer name, incidents occurrence, weather conditions affecting flight, etc), as well as instructor/examiner signature when appropriate, in this section.

FLT. NO.	DATE		AIRCRAFT	FLIGHT LOCATION			FLIGHT DURATION				FUNCTION			DATA	
	Yr ____		Make/Model	Takeoff		Landing	Fixed Wing	Rotary	LTA	Total	OIC/Solo	Training Received	Instructor	Ld.	Dist.
	Dd / Mm		Registration No.	Time		Time									
	__ / __														
	__ / __														
	__ / __														
	__ / __														
	__ / __														
	__ / __														
	__ / __														
	__ / __														
I certify that the entries in this log are true.				Totals this page:											
				Totals brought forward:											
				Totals to date:											

Flt. No.	Batteries / Fuel	P-f. Ins. C. Cal.	Technical Problems Found / Corrective Action, Status	Remarks and Endorsements

FLT. NO.	DATE	AIRCRAFT	FLIGHT LOCATION		FLIGHT DURATION				FUNCTION			DATA	
	Yr ____	Make/Model	Takeoff	Landing	Fixed Wing	Rotary	LTA	Total	OIC/Solo	Training Received	Instructor	Ld.	Dist.
	Dd / Mm	Registration No.	Time	Time									
	__/__												
	__/__												
	__/__												
	__/__												
	__/__												
	__/__												
	__/__												
	__/__												

I certify that the entries in this log are true.	Totals this page:								
	Totals brought forward:								
	Totals to date:								

Flt. No.	Batteries / Fuel	P-f. Ins. C. Cal.	Technical Problems Found / Corrective Action, Status	Remarks and Endorsements

23

FLT. NO.	DATE Yr ____ Dd / Mm	AIRCRAFT Make/Model Registration No.	FLIGHT LOCATION		FLIGHT DURATION				FUNCTION			DATA	
			Takeoff Time	Landing Time	Fixed Wing	Rotary	LTA	Total	OIC/Solo	Training Received	Instructor	Ld.	Dist.
	__/__												
	__/__												
	__/__												
	__/__												
	__/__												
	__/__												
	__/__												
	__/__												

I certify that the entries in this log are true.

Totals this page:			
Totals brought forward:			
Totals to date:			

24

Flt. No.	Batteries / Fuel	P-f. Ins. C. Cal.	Technical Problems Found / Corrective Action, Status	Remarks and Endorsements

NOTES CONCERNING THE ABOVE FLIGHTS

FLT. NO.	DATE Yr ____ Dd / Mm	AIRCRAFT Make/Model Registration No.	FLIGHT LOCATION Takeoff Time	FLIGHT LOCATION Landing Time	FLIGHT DURATION Fixed Wing	Rotary	LTA	Total	FUNCTION OIC/Solo	Training Received	Instructor	DATA Ld.	Dist.
	__/__												
	__/__												
	__/__												
	__/__												
	__/__												
	__/__												
	__/__												
	__/__												

I certify that the entries in this log are true.

Totals this page:	
Totals brought forward:	
Totals to date:	

Flt. No.	Batteries / Fuel	P-f. Ins. C. Cal.	Technical Problems Found / Corrective Action, Status	Remarks and Endorsements

NOTES CONCERNING THE ABOVE FLIGHTS

FLT. No.	DATE	AIRCRAFT	FLIGHT LOCATION		FLIGHT DURATION				FUNCTION			DATA	
	Yr ____	Make/Model	Takeoff	Landing	Fixed Wing	Rotary	LTA	Total	OIC/Solo	Training Received	Instructor	Ld.	Dist.
	Dd / Mm	Registration No.	Time	Time									
	__/__												
	__/__												
	__/__												
	__/__												
	__/__												
	__/__												
	__/__												
	__/__												

I certify that the entries in this log are true.	Totals this page:									
	Totals brought forward:									
	Totals to date:									

Flt. No.	Batteries / Fuel	P-f. Ins. C. Cal.	Technical Problems Found / Corrective Action, Status	Remarks and Endorsements

FLT. NO.	DATE Yr ____	AIRCRAFT	FLIGHT LOCATION		FLIGHT DURATION				FUNCTION			DATA	
		Make/Model	Takeoff	Landing	Fixed Wing	Rotary	LTA	Total	OIC/Solo	Training Received	Instructor	Ld.	Dist.
	Dd / Mm	Registration No.	Time	Time									
	__/__												
	__/__												
	__/__												
	__/__												
	__/__												
	__/__												
	__/__												
	__/__												
I certify that the entries in this log are true.				Totals this page:									
				Totals brought forward:									
				Totals to date:									

			NOTES CONCERNING THE ABOVE FLIGHTS	
Flt. No.	Batteries / Fuel	P-f. Ins. C. Cal.	Technical Problems Found / Corrective Action, Status	Remarks and Endorsements

Flt. No.	Date	Aircraft	Flight Location		Flight Duration				Function			Data	
	Yr ____	Make/Model	Takeoff	Landing	Fixed Wing	Rotary	LTA	Total	OIC/Solo	Training Received	Instructor	Ld.	Dist.
	Dd / Mm	Registration No.	Time	Time									
	__ / __												
	__ / __												
	__ / __												
	__ / __												
	__ / __												
	__ / __												
	__ / __												
	__ / __												
I certify that the entries in this log are true.			Totals this page:										
			Totals brought forward:										
			Totals to date:										

Flt. No.	Batteries / Fuel	P-f. Ins. C. Cal.	Technical Problems Found / Corrective Action, Status	Remarks and Endorsements

NOTES CONCERNING THE ABOVE FLIGHTS

FLT. NO.	DATE	AIRCRAFT	FLIGHT LOCATION		FLIGHT DURATION				FUNCTION			DATA	
	Yr ____	Make/Model	Takeoff	Landing	Fixed Wing	Rotary	LTA	Total	OIC/Solo	Training Received	Instructor	Ld.	Dist.
	Dd / Mm	Registration No.	Time	Time									
	__/__												
	__/__												
	__/__												
	__/__												
	__/__												
	__/__												
	__/__												
	__/__												
I certify that the entries in this log are true.			Totals this page:										
			Totals brought forward:										
			Totals to date:										

Flt. No.	Batteries / Fuel	P-f. Ins. C. Cal.	Technical Problems Found / Corrective Action, Status	Remarks and Endorsements

FLT. NO.	DATE	AIRCRAFT	FLIGHT LOCATION		FLIGHT DURATION				FUNCTION			DATA	
	Yr ____	Make/Model	Takeoff	Landing	Fixed Wing	Rotary	LTA	Total	OIC/Solo	Training Received	Instructor	Ld.	Dist.
	Dd / Mm	Registration No.	Time	Time									
	__/__												
	__/__												
	__/__												
	__/__												
	__/__												
	__/__												
	__/__												
	__/__												
I certify that the entries in this log are true.			Totals this page:										
			Totals brought forward:										
			Totals to date:										

Flt. No.	Batteries / Fuel	P-f. Ins. C. Cal.	Technical Problems Found / Corrective Action, Status	Remarks and Endorsements

NOTES CONCERNING THE ABOVE FLIGHTS

Flt. No.	Date Yr ___ Dd / Mm	Aircraft Make/Model Registration No.	Flight Location		Flight Duration				Function			Data	
			Takeoff Time	Landing Time	Fixed Wing	Rotary	LTA	Total	OIC/Solo	Training Received	Instructor	Ld.	Dist.
	__/__												
	__/__												
	__/__												
	__/__												
	__/__												
	__/__												
	__/__												
	__/__												

I certify that the entries in this log are true.

Totals this page:		
Totals brought forward:		
Totals to date:		

			NOTES CONCERNING THE ABOVE FLIGHTS	
Flt. No.	Batteries / Fuel	P-f. Ins. C. Cal.	Technical Problems Found / Corrective Action, Status	Remarks and Endorsements

FLT. NO.	DATE	AIRCRAFT	FLIGHT LOCATION		FLIGHT DURATION				FUNCTION			DATA	
	Yr ____	Make/Model	Takeoff	Landing	Fixed Wing	Rotary	LTA	Total	OIC/Solo	Training Received	Instructor	Ld.	Dist.
	Dd / Mm	Registration No.	Time	Time									
	__/__												
	__/__												
	__/__												
	__/__												
	__/__												
	__/__												
	__/__												
	__/__												
I certify that the entries in this log are true.			Totals this page:										
			Totals brought forward:										
			Totals to date:										

Flt. No.	Batteries / Fuel	P-f. Ins. C. Cal.	Technical Problems Found / Corrective Action, Status	Remarks and Endorsements

NOTES CONCERNING THE ABOVE FLIGHTS

Flt. No.	Date		Aircraft	Flight Location			Flight Duration				Function			Data	
	Yr ____		Make/Model	Takeoff	Landing		Fixed Wing	Rotary	LTA	Total	OIC/Solo	Training Received	Instructor	Ld.	Dist.
	Dd / Mm		Registration No.	Time	Time										
	__/__														
	__/__														
	__/__														
	__/__														
	__/__														
	__/__														
	__/__														
	__/__														
I certify that the entries in this log are true.					Totals this page:										
					Totals brought forward:										
					Totals to date:										

Flt. No.	Batteries / Fuel	P-f. Ins. C. Cal.	Technical Problems Found / Corrective Action, Status	Remarks and Endorsements

NOTES CONCERNING THE ABOVE FLIGHTS

FLT. NO.	DATE	AIRCRAFT	FLIGHT LOCATION		FLIGHT DURATION				FUNCTION			DATA	
	Yr ____	Make/Model	Takeoff	Landing	Fixed Wing	Rotary	LTA	Total	OIC/Solo	Training Received	Instructor	Ld.	Dist.
	Dd / Mm	Registration No.	Time	Time									
	__/__												
	__/__												
	__/__												
	__/__												
	__/__												
	__/__												
	__/__												
	__/__												

I certify that the entries in this log are true.	Totals this page:								
	Totals brought forward:								
	Totals to date:								

NOTES CONCERNING THE ABOVE FLIGHTS				
Flt. No.	Batteries / Fuel	P-f. Ins. C. Cal.	Technical Problems Found / Corrective Action, Status	Remarks and Endorsements

FLT. NO.	DATE	AIRCRAFT	FLIGHT LOCATION		FLIGHT DURATION				FUNCTION			DATA	
	Yr ____	Make/Model	Takeoff	Landing	Fixed Wing	Rotary	LTA	Total	OIC/Solo	Training Received	Instructor	Ld.	Dist.
	Dd / Mm	Registration No.	Time	Time									
	__/__												
	__/__												
	__/__												
	__/__												
	__/__												
	__/__												
	__/__												
	__/__												
I certify that the entries in this log are true.				Totals this page:									
				Totals brought forward:									
				Totals to date:									

Flt. No.	Batteries / Fuel	P-f. Ins. C. Cal.	Technical Problems Found / Corrective Action, Status	Remarks and Endorsements

NOTES CONCERNING THE ABOVE FLIGHTS

FLT. NO.	DATE	AIRCRAFT	FLIGHT LOCATION		FLIGHT DURATION				FUNCTION			DATA	
	Yr ____	Make/Model	Takeoff	Landing	Fixed Wing	Rotary	LTA	Total	OIC/Solo	Training Received	Instructor	Ld.	Dist.
	Dd / Mm	Registration No.	Time	Time									
	__/__												
	__/__												
	__/__												
	__/__												
	__/__												
	__/__												
	__/__												
	__/__												
I certify that the entries in this log are true.			Totals this page:										
			Totals brought forward:										
			Totals to date:										

Flt. No.	Batteries / Fuel	P-f. Ins. C. Cal.	Technical Problems Found / Corrective Action, Status	Remarks and Endorsements
NOTES CONCERNING THE ABOVE FLIGHTS				

Flt. No.	Date Yr ___ Dd / Mm	Aircraft Make/Model Registration No.	Flight Location		Flight Duration				Function			Data	
			Takeoff Time	Landing Time	Fixed Wing	Rotary	LTA	Total	OIC/Solo	Training Received	Instructor	Ld.	Dist.
	__/__												
	__/__												
	__/__												
	__/__												
	__/__												
	__/__												
	__/__												
	__/__												
I certify that the entries in this log are true.			Totals this page:										
			Totals brought forward:										
			Totals to date:										

Flt. No.	Batteries / Fuel	P-f. Ins. C. Cal.	Technical Problems Found / Corrective Action, Status	Remarks and Endorsements

FLT. NO.	DATE		AIRCRAFT	FLIGHT LOCATION			FLIGHT DURATION				FUNCTION			DATA	
	Yr ____		Make/Model	Takeoff	Landing		Fixed Wing	Rotary	LTA	Total	OIC/Solo	Training Received	Instructor	Ld.	Dist.
	Dd / Mm		Registration No.	Time	Time										
	__ / __														
	__ / __														
	__ / __														
	__ / __														
	__ / __														
	__ / __														
	__ / __														
	__ / __														
I certify that the entries in this log are true.					Totals this page:										
					Totals brought forward:										
					Totals to date:										

Flt. No.	Batteries / Fuel	P-f. Ins. C. Cal.	Technical Problems Found / Corrective Action, Status	Remarks and Endorsements

FLT. NO.	DATE	AIRCRAFT	FLIGHT LOCATION		FLIGHT DURATION				FUNCTION			DATA	
	Yr ___	Make/Model	Takeoff	Landing	Fixed Wing	Rotary	LTA	Total	OIC/Solo	Training Received	Instructor	Ld.	Dist.
	Dd / Mm	Registration No.	Time	Time									
	__/__												
	__/__												
	__/__												
	__/__												
	__/__												
	__/__												
	__/__												
	__/__												
I certify that the entries in this log are true.				Totals this page:									
				Totals brought forward:									
				Totals to date:									

54

Flt. No.	Batteries / Fuel	P-f. Ins. C. Cal.	Technical Problems Found / Corrective Action, Status	Remarks and Endorsements
NOTES CONCERNING THE ABOVE FLIGHTS				

Flt. No.	Date	Aircraft	Flight Location		Flight Duration				Function			Data	
	Yr ____	Make/Model	Takeoff	Landing	Fixed Wing	Rotary	LTA	Total	OIC/Solo	Training Received	Instructor	Ld.	Dist.
	Dd / Mm	Registration No.	Time	Time									
	__/__												
	__/__												
	__/__												
	__/__												
	__/__												
	__/__												
	__/__												
	__/__												

I certify that the entries in this log are true.

Totals this page:		
Totals brought forward:		
Totals to date:		

Flt. No.	Batteries / Fuel	P-f. Ins. C. Cal.	Technical Problems Found / Corrective Action, Status	Remarks and Endorsements
NOTES CONCERNING THE ABOVE FLIGHTS				

FLT. NO.	DATE Yr ____ Dd / Mm	AIRCRAFT Make/Model Registration No.	FLIGHT LOCATION		FLIGHT DURATION				FUNCTION			DATA	
			Takeoff Time	Landing Time	Fixed Wing	Rotary	LTA	Total	OIC/Solo	Training Received	Instructor	Ld.	Dist.
	__/__												
	__/__												
	__/__												
	__/__												
	__/__												
	__/__												
	__/__												
	__/__												

I certify that the entries in this log are true.		Totals this page:								
		Totals brought forward:								
		Totals to date:								

Flt. No.	Batteries / Fuel	P-f. Ins. C. Cal.	Technical Problems Found / Corrective Action, Status	Remarks and Endorsements
NOTES CONCERNING THE ABOVE FLIGHTS				

FLT. NO.	DATE	AIRCRAFT	FLIGHT LOCATION		FLIGHT DURATION				FUNCTION			DATA	
	Yr ____	Make/Model	Takeoff	Landing	Fixed Wing	Rotary	LTA	Total	OIC/Solo	Training Received	Instructor	Ld.	Dist.
	Dd / Mm	Registration No.	Time	Time									
	__/__												
	__/__												
	__/__												
	__/__												
	__/__												
	__/__												
	__/__												
	__/__												
I certify that the entries in this log are true.			Totals this page:										
			Totals brought forward:										
			Totals to date:										

Flt. No.	Batteries / Fuel	P-f. Ins. C. Cal.	Technical Problems Found / Corrective Action, Status	Remarks and Endorsements

FLT. NO.	DATE	AIRCRAFT	FLIGHT LOCATION		FLIGHT DURATION				FUNCTION			DATA	
	Yr ____	Make/Model	Takeoff	Landing	Fixed Wing	Rotary	LTA	Total	OIC/Solo	Training Received	Instructor	Ld.	Dist.
	Dd / Mm	Registration No.	Time	Time									
	__/__												
	__/__												
	__/__												
	__/__												
	__/__												
	__/__												
	__/__												
	__/__												
I certify that the entries in this log are true.			Totals this page:										
			Totals brought forward:										
			Totals to date:										

Flt. No.	Batteries / Fuel	P-f. Ins. C. Cal.	Technical Problems Found / Corrective Action, Status	Remarks and Endorsements

FLT. NO.	DATE	AIRCRAFT	FLIGHT LOCATION		FLIGHT DURATION				FUNCTION			DATA	
	Yr ____	Make/Model	Takeoff	Landing	Fixed Wing	Rotary	LTA	Total	OIC/Solo	Training Received	Instructor	Ld.	Dist.
	Dd / Mm	Registration No.	Time	Time									
	__/__												
	__/__												
	__/__												
	__/__												
	__/__												
	__/__												
	__/__												
	__/__												
I certify that the entries in this log are true.			Totals this page:										
			Totals brought forward:										
			Totals to date:										

NOTES CONCERNING THE ABOVE FLIGHTS

Flt. No.	Batteries / Fuel	P-f. Ins. C. Cal.	Technical Problems Found / Corrective Action, Status	Remarks and Endorsements

FLT. NO.	DATE Yr ____ Dd / Mm	AIRCRAFT Make/Model Registration No.	FLIGHT LOCATION Takeoff — Time	Landing — Time	FLIGHT DURATION Fixed Wing	Rotary	LTA	Total	FUNCTION OIC/Solo	Training Received	Instructor	DATA Ld.	Dist.
	__/__												
	__/__												
	__/__												
	__/__												
	__/__												
	__/__												
	__/__												
	__/__												
I certify that the entries in this log are true.			Totals this page:										
			Totals brought forward:										
			Totals to date:										

Flt. No.	Batteries / Fuel	P-f. Ins. C. Cal.	Technical Problems Found / Corrective Action, Status	Remarks and Endorsements

FLT. NO.	DATE	AIRCRAFT		FLIGHT LOCATION			FLIGHT DURATION				FUNCTION			DATA	
	Yr ____	Make/Model		Takeoff	Landing		Fixed Wing	Rotary	LTA	Total	OIC/Solo	Training Received	Instructor	Ld.	Dist.
	Dd / Mm	Registration No.		Time	Time										
	__/__														
	__/__														
	__/__														
	__/__														
	__/__														
	__/__														
	__/__														
	__/__														

I certify that the entries in this log are true.

Totals this page:						
Totals brought forward:						
Totals to date:						

Flt. No.	Batteries / Fuel	P-f. Ins. C. Cal.	Technical Problems Found / Corrective Action, Status	Remarks and Endorsements
NOTES CONCERNING THE ABOVE FLIGHTS				

FLT. NO.	DATE Yr ____ Dd / Mm	AIRCRAFT Make/Model Registration No.	FLIGHT LOCATION Takeoff Time	Landing Time	FLIGHT DURATION Fixed Wing	Rotary	LTA	Total	FUNCTION OIC/Solo	Training Received	Instructor	DATA Ld.	Dist.
	__ / __												
	__ / __												
	__ / __												
	__ / __												
	__ / __												
	__ / __												
	__ / __												
	__ / __												

I certify that the entries in this log are true.	Totals this page:								
	Totals brought forward:								
	Totals to date:								

	Notes concerning the above flights			
Flt. No.	Batteries / Fuel	P-f. Ins. C. Cal.	Technical Problems Found / Corrective Action, Status	Remarks and Endorsements

Flt. No.	Date	Aircraft	Flight Location		Flight Duration				Function			Data	
	Yr ____	Make/Model	Takeoff	Landing	Fixed Wing	Rotary	LTA	Total	OIC/Solo	Training Received	Instructor	Ld.	Dist.
	Dd / Mm	Registration No.	Time	Time									
	___/___												
	___/___												
	___/___												
	___/___												
	___/___												
	___/___												
	___/___												
	___/___												
I certify that the entries in this log are true.			Totals this page:										
			Totals brought forward:										
			Totals to date:										

Flt. No.	Batteries / Fuel	P-f. Ins. C. Cal.	Technical Problems Found / Corrective Action, Status	Remarks and Endorsements

73

Flt. No.	Date Yr ____ Dd / Mm	Aircraft Make/Model Registration No.	Flight Location		Flight Duration				Function			Data	
			Takeoff Time	Landing Time	Fixed Wing	Rotary	LTA	Total	OIC/Solo	Training Received	Instructor	Ld.	Dist.
	__/__												
	__/__												
	__/__												
	__/__												
	__/__												
	__/__												
	__/__												
	__/__												

I certify that the entries in this log are true.

Totals this page:			
Totals brought forward:			
Totals to date:			

Flt. No.	Batteries / Fuel	P-f. Ins. C. Cal.	Technical Problems Found / Corrective Action, Status	Remarks and Endorsements

NOTES CONCERNING THE ABOVE FLIGHTS

FLT. NO.	DATE	AIRCRAFT	FLIGHT LOCATION		FLIGHT DURATION				FUNCTION			DATA	
	Yr ____	Make/Model	Takeoff	Landing	Fixed Wing	Rotary	LTA	Total	OIC/Solo	Training Received	Instructor	Ld.	Dist.
	Dd / Mm	Registration No.	Time	Time									
	__/__												
	__/__												
	__/__												
	__/__												
	__/__												
	__/__												
	__/__												
	__/__												
I certify that the entries in this log are true.			Totals this page:										
			Totals brought forward:										
			Totals to date:										

Flt. No.	Batteries / Fuel	P-f. Ins. C. Cal.	Technical Problems Found / Corrective Action, Status	Remarks and Endorsements
NOTES CONCERNING THE ABOVE FLIGHTS				

FLT. NO.	DATE Yr ___ Dd / Mm	AIRCRAFT Make/Model Registration No.	FLIGHT LOCATION Takeoff Time	Landing Time	FLIGHT DURATION Fixed Wing	Rotary	LTA	Total	FUNCTION OIC/Solo	Training Received	Instructor	DATA Ld.	Dist.
	__ / __												
	__ / __												
	__ / __												
	__ / __												
	__ / __												
	__ / __												
	__ / __												
	__ / __												
I certify that the entries in this log are true.			Totals this page:										
			Totals brought forward:										
			Totals to date:										

Flt. No.	Batteries / Fuel	P-f. Ins. C. Cal.	Technical Problems Found / Corrective Action, Status	Remarks and Endorsements

Flt. No.	Date	Aircraft	Flight Location		Flight Duration				Function			Data	
	Yr ____	Make/Model	Takeoff	Landing	Fixed Wing	Rotary	LTA	Total	OIC/Solo	Training Received	Instructor	Ld.	Dist.
	Dd / Mm	Registration No.	Time	Time									
	__/__												
	__/__												
	__/__												
	__/__												
	__/__												
	__/__												
	__/__												
	__/__												

I certify that the entries in this log are true.

Totals this page:			
Totals brought forward:			
Totals to date:			

NOTES CONCERNING THE ABOVE FLIGHTS

Flt. No.	Batteries / Fuel	P-f. Ins. C. Cal.	Technical Problems Found / Corrective Action, Status	Remarks and Endorsements

FLT. NO.	DATE Yr ____ Dd / Mm	AIRCRAFT Make/Model Registration No.	FLIGHT LOCATION Takeoff Time	FLIGHT LOCATION Landing Time	FLIGHT DURATION Fixed Wing	FLIGHT DURATION Rotary	FLIGHT DURATION LTA	FLIGHT DURATION Total	FUNCTION OIC/Solo	FUNCTION Training Received	FUNCTION Instructor	DATA Ld.	DATA Dist.
	__/__												
	__/__												
	__/__												
	__/__												
	__/__												
	__/__												
	__/__												
	__/__												
I certify that the entries in this log are true.				Totals this page:									
				Totals brought forward:									
				Totals to date:									

NOTES CONCERNING THE ABOVE FLIGHTS

Flt. No.	Batteries / Fuel	P-f. Ins. C. Cal.	Technical Problems Found / Corrective Action, Status	Remarks and Endorsements

FLT. NO.	DATE	AIRCRAFT	FLIGHT LOCATION		FLIGHT DURATION				FUNCTION			DATA	
	Yr ____	Make/Model	Takeoff	Landing	Fixed Wing	Rotary	LTA	Total	OIC/Solo	Training Received	Instructor	Ld.	Dist.
	Dd / Mm	Registration No.	Time	Time									
	__/__												
	__/__												
	__/__												
	__/__												
	__/__												
	__/__												
	__/__												
	__/__												

I certify that the entries in this log are true.

Totals this page:		
Totals brought forward:		
Totals to date:		

Flt. No.	Batteries / Fuel	P-f. Ins. C. Cal.	Technical Problems Found / Corrective Action, Status	Remarks and Endorsements

FLT. NO.	DATE Yr ____ Dd / Mm	AIRCRAFT Make/Model Registration No.	FLIGHT LOCATION Takeoff / Time	FLIGHT LOCATION Landing / Time	FLIGHT DURATION Fixed Wing	Rotary	LTA	Total	FUNCTION OIC/Solo	Training Received	Instructor	DATA Ld.	Dist.
	__ / __												
	__ / __												
	__ / __												
	__ / __												
	__ / __												
	__ / __												
	__ / __												
	__ /												
I certify that the entries in this log are true.				Totals this page:									
				Totals brought forward:									
				Totals to date:									

Flt. No.	Batteries / Fuel	P-f. Ins. C. Cal.	Technical Problems Found / Corrective Action, Status	Remarks and Endorsements
NOTES CONCERNING THE ABOVE FLIGHTS				

FLT. NO.	DATE Yr ____ Dd / Mm	AIRCRAFT Make/Model Registration No.	FLIGHT LOCATION Takeoff Time	Landing Time	FLIGHT DURATION Fixed Wing	Rotary	LTA	Total	FUNCTION OIC/Solo	Training Received	Instructor	DATA Ld.	Dist.
	__ / __												
	__ / __												
	__ / __												
	__ / __												
	__ / __												
	__ / __												
	__ / __												
	__ / __												

I certify that the entries in this log are true.

Totals this page:			
Totals brought forward:			
Totals to date:			

NOTES CONCERNING THE ABOVE FLIGHTS

Flt. No.	Batteries / Fuel	P-f. Ins. C. Cal.	Technical Problems Found / Corrective Action, Status	Remarks and Endorsements

FLT. NO.	DATE Yr ____ Dd / Mm	AIRCRAFT Make/Model Registration No.	FLIGHT LOCATION		FLIGHT DURATION				FUNCTION			DATA	
			Takeoff Time	Landing Time	Fixed Wing	Rotary	LTA	Total	OIC/Solo	Training Received	Instructor	Ld.	Dist.
	__/__												
	__/__												
	__/__												
	__/__												
	__/__												
	__/__												
	__/__												
	__/__												
I certify that the entries in this log are true.				Totals this page:									
				Totals brought forward:									
				Totals to date:									

Flt. No.	Batteries / Fuel	P-f. Ins. C. Cal.	Technical Problems Found / Corrective Action, Status	Remarks and Endorsements

FLT. NO.	DATE	AIRCRAFT	FLIGHT LOCATION		FLIGHT DURATION				FUNCTION			DATA	
	Yr ____	Make/Model	Takeoff	Landing	Fixed Wing	Rotary	LTA	Total	OIC/Solo	Training Received	Instructor	Ld.	Dist.
	Dd / Mm	Registration No.	Time	Time									
	__/__												
	__/__												
	__/__												
	__/__												
	__/__												
	__/__												
	__/__												
	__/__												

I certify that the entries in this log are true.

Totals this page:		
Totals brought forward:		
Totals to date:		

Flt. No.	Batteries / Fuel	P-f. Ins. C. Cal.	Technical Problems Found / Corrective Action, Status	Remarks and Endorsements

FLT. NO.	DATE Yr ____ Dd / Mm	AIRCRAFT Make/Model Registration No.	FLIGHT LOCATION		FLIGHT DURATION				FUNCTION			DATA	
			Takeoff Time	Landing Time	Fixed Wing	Rotary	LTA	Total	OIC/Solo	Training Received	Instructor	Ld.	Dist.
	__/__												
	__/__												
	__/__												
	__/__												
	__/__												
	__/__												
	__/__												
	__/__												

I certify that the entries in this log are true.

Totals this page:		
Totals brought forward:		
Totals to date:		

NOTES CONCERNING THE ABOVE FLIGHTS

Flt. No.	Batteries / Fuel	P-f. Ins. C. Cal.	Technical Problems Found / Corrective Action, Status	Remarks and Endorsements

FLT. NO.	DATE Yr ___ Dd / Mm	AIRCRAFT Make/Model Registration No.	FLIGHT LOCATION Takeoff Time	Landing Time	FLIGHT DURATION Fixed Wing	Rotary	LTA	Total	FUNCTION OIC/Solo	Training Received	Instructor	DATA Ld.	Dist.
	__ / __												
	__ / __												
	__ / __												
	__ / __												
	__ / __												
	__ / __												
	__ / __												
	__ / __												
I certify that the entries in this log are true.			Totals this page:										
			Totals brought forward:										
			Totals to date:										

NOTES CONCERNING THE ABOVE FLIGHTS

Flt. No.	Batteries / Fuel	P-f. Ins. C. Cal.	Technical Problems Found / Corrective Action, Status	Remarks and Endorsements

FLT. NO.	DATE Yr ____ Dd / Mm	AIRCRAFT Make/Model Registration No.	FLIGHT LOCATION		FLIGHT DURATION				FUNCTION			DATA	
			Takeoff Time	Landing Time	Fixed Wing	Rotary	LTA	Total	OIC/Solo	Training Received	Instructor	Ld.	Dist.
	__/__												
	__/__												
	__/__												
	__/__												
	__/__												
	__/__												
	__/__												
	__/__												
I certify that the entries in this log are true.			Totals this page:										
			Totals brought forward:										
			Totals to date:										

NOTES CONCERNING THE ABOVE FLIGHTS

Flt. No.	Batteries / Fuel	P-f. Ins. C. Cal.	Technical Problems Found / Corrective Action, Status	Remarks and Endorsements

FLT. NO.	DATE	AIRCRAFT		FLIGHT LOCATION		FLIGHT DURATION				FUNCTION			DATA	
	Yr ____	Make/Model		Takeoff	Landing	Fixed Wing	Rotary	LTA	Total	OIC/Solo	Training Received	Instructor	Ld.	Dist.
	Dd / Mm	Registration No.		Time	Time									
	__/__													
	__/__													
	__/__													
	__/__													
	__/__													
	__/__													
	__/__													
	__/__													

I certify that the entries in this log are true.

	Totals this page:								
Totals brought forward:									
Totals to date:									

NOTES CONCERNING THE ABOVE FLIGHTS

Flt. No.	Batteries / Fuel	P-f. Ins. C. Cal.	Technical Problems Found / Corrective Action, Status	Remarks and Endorsements

FLT. NO.	DATE Yr ____ Dd / Mm	AIRCRAFT Make/Model Registration No.	FLIGHT LOCATION Takeoff Time	Landing Time	FLIGHT DURATION Fixed Wing	Rotary	LTA	Total	FUNCTION OIC/Solo	Training Received	Instructor	DATA Ld.	Dist.
	__/__												
	__/__												
	__/__												
	__/__												
	__/__												
	__/__												
	__/__												
	__/__												

I certify that the entries in this log are true.

Totals this page:

Totals brought forward:

Totals to date:

NOTES CONCERNING THE ABOVE FLIGHTS

Flt. No.	Batteries / Fuel	P-f. Ins. C. Cal.	Technical Problems Found / Corrective Action, Status	Remarks and Endorsements

FLT. NO.	DATE Yr ____ Dd / Mm	AIRCRAFT Make/Model Registration No.	FLIGHT LOCATION		FLIGHT DURATION				FUNCTION			DATA	
			Takeoff Time	Landing Time	Fixed Wing	Rotary	LTA	Total	OIC/Solo	Training Received	Instructor	Ld.	Dist.
	__ / __												
	__ / __												
	__ / __												
	__ / __												
	__ / __												
	__ / __												
	__ / __												
	__ / __												
I certify that the entries in this log are true.				Totals this page:									
				Totals brought forward:									
				Totals to date:									

NOTES CONCERNING THE ABOVE FLIGHTS

Flt. No.	Batteries / Fuel	P-f. Ins. C. Cal.	Technical Problems Found / Corrective Action, Status	Remarks and Endorsements

FLT. NO.	DATE Yr ____ Dd / Mm	AIRCRAFT Make/Model Registration No.	FLIGHT LOCATION Takeoff Time	Landing Time	FLIGHT DURATION Fixed Wing	Rotary	LTA	Total	FUNCTION OIC/Solo	Training Received	Instructor	DATA Ld.	Dist.
	__ / __												
	__ / __												
	__ / __												
	__ / __												
	__ / __												
	__ / __												
	__ / __												
	__ / __												

I certify that the entries in this log are true.	Totals this page:									
	Totals brought forward:									
	Totals to date:									

NOTES CONCERNING THE ABOVE FLIGHTS

Flt. No.	Batteries / Fuel	P-f. Ins. C. Cal.	Technical Problems Found / Corrective Action, Status	Remarks and Endorsements

FLT. NO.	DATE		AIRCRAFT	FLIGHT LOCATION			FLIGHT DURATION				FUNCTION			DATA	
	Yr ____		Make/Model	Takeoff		Landing	Fixed Wing	Rotary	LTA	Total	OIC/Solo	Training Received	Instructor	Ld.	Dist.
	Dd / Mm		Registration No.	Time		Time									
	__ / __														
	__ / __														
	__ / __														
	__ / __														
	__ / __														
	__ / __														
	__ / __														
	__ / __														
I certify that the entries in this log are true.				Totals this page:											
				Totals brought forward:											
				Totals to date:											

NOTES CONCERNING THE ABOVE FLIGHTS

Flt. No.	Batteries / Fuel	P-f. Ins. C. Cal.	Technical Problems Found / Corrective Action, Status	Remarks and Endorsements

FLT. NO.	DATE	AIRCRAFT	FLIGHT LOCATION		FLIGHT DURATION				FUNCTION			DATA	
	Yr ____	Make/Model	Takeoff	Landing	Fixed Wing	Rotary	LTA	Total	OIC/Solo	Training Received	Instructor	Ld.	Dist.
	Dd / Mm	Registration No.	Time	Time									
	__ / __												
	__ / __												
	__ / __												
	__ / __												
	__ / __												
	__ / __												
	__ / __												
	__ / __												
I certify that the entries in this log are true.				Totals this page:									
				Totals brought forward:									
				Totals to date:									

NOTES CONCERNING THE ABOVE FLIGHTS

Flt. No.	Batteries / Fuel	P-f. Ins. C. Cal.	Technical Problems Found / Corrective Action, Status	Remarks and Endorsements

FLT. NO.	DATE	AIRCRAFT	FLIGHT LOCATION		FLIGHT DURATION				FUNCTION			DATA	
	Yr ___	Make/Model	Takeoff	Landing	Fixed Wing	Rotary	LTA	Total	OIC/Solo	Training Received	Instructor	Ld.	Dist.
	Dd / Mm	Registration No.	Time	Time									
	__ / __												
	__ / __												
	__ / __												
	__ / __												
	__ / __												
	__ / __												
	__ / __												
	__ / __												

I certify that the entries in this log are true.

Totals this page:		
Totals brought forward:		
Totals to date:		

NOTES CONCERNING THE ABOVE FLIGHTS

Flt. No.	Batteries / Fuel	P-f. Ins. C. Cal.	Technical Problems Found / Corrective Action, Status	Remarks and Endorsements

Flt. No.	Date Yr ___ Dd / Mm	Aircraft Make/Model Registration No.	Flight Location		Flight Duration				Function			Data	
			Takeoff Time	Landing Time	Fixed Wing	Rotary	LTA	Total	OIC/Solo	Training Received	Instructor	Ld.	Dist.
	__/__												
	__/__												
	__/__												
	__/__												
	__/__												
	__/__												
	__/__												
	__/												

I certify that the entries in this log are true.

Totals this page:		
Totals brought forward:		
Totals to date:		

Flt. No.	Batteries / Fuel	P-f. Ins. C. Cal.	Technical Problems Found / Corrective Action, Status	Remarks and Endorsements

FLT. NO.	DATE Yr ____ Dd / Mm	AIRCRAFT Make/Model Registration No.	FLIGHT LOCATION Takeoff Time	Landing Time	FLIGHT DURATION Fixed Wing	Rotary	LTA	Total	FUNCTION OIC/Solo	Training Received	Instructor	DATA Ld.	Dist.
	__/__												
	__/__												
	__/__												
	__/__												
	__/__												
	__/__												
	__/__												
	__/__												

I certify that the entries in this log are true.

Totals this page:

Totals brought forward:

Totals to date:

Flt. No.	Batteries / Fuel	P-f. Ins. C. Cal.	Technical Problems Found / Corrective Action, Status	Remarks and Endorsements

FLT. NO.	DATE Yr ____ Dd / Mm	AIRCRAFT Make/Model Registration No.	FLIGHT LOCATION Takeoff Time	Landing Time	FLIGHT DURATION Fixed Wing	Rotary	LTA	Total	FUNCTION OIC/Solo	Training Received	Instructor	DATA Ld.	Dist.
	__/__												
	__/__												
	__/__												
	__/__												
	__/__												
	__/__												
	__/__												
	__/__												
I certify that the entries in this log are true.			Totals this page:										
			Totals brought forward:										
			Totals to date:										

NOTES CONCERNING THE ABOVE FLIGHTS

Flt. No.	Batteries / Fuel	P-f. Ins. C. Cal.	Technical Problems Found / Corrective Action, Status	Remarks and Endorsements

FLT. NO.	DATE	AIRCRAFT	FLIGHT LOCATION		FLIGHT DURATION				FUNCTION			DATA	
	Yr ____	Make/Model	Takeoff	Landing	Fixed Wing	Rotary	LTA	Total	OIC/Solo	Training Received	Instructor	Ld.	Dist.
	Dd / Mm	Registration No.	Time	Time									
	__/__												
	__/__												
	__/__												
	__/__												
	__/__												
	__/__												
	__/__												
	__/__												

I certify that the entries in this log are true.

Totals this page:		
Totals brought forward:		
Totals to date:		

Flt. No.	Batteries / Fuel	P-f. Ins. C. Cal.	Technical Problems Found / Corrective Action, Status	Remarks and Endorsements

FLT. NO.	DATE	AIRCRAFT	FLIGHT LOCATION		FLIGHT DURATION				FUNCTION			DATA	
	Yr ___	Make/Model	Takeoff	Landing	Fixed Wing	Rotary	LTA	Total	OIC/Solo	Training Received	Instructor	Ld.	Dist.
	Dd / Mm	Registration No.	Time	Time									
	__/__												
	__/__												
	__/__												
	__/__												
	__/__												
	__/__												
	__/__												
	__/__												

I certify that the entries in this log are true.		Totals this page:								
		Totals brought forward:								
		Totals to date:								

Flt. No.	Batteries / Fuel	P-f. Ins. C. Cal.	Technical Problems Found / Corrective Action, Status	Remarks and Endorsements

123

FLT. NO.	DATE		AIRCRAFT	FLIGHT LOCATION			FLIGHT DURATION				FUNCTION			DATA	
	Yr ___		Make/Model	Takeoff		Landing	Fixed Wing	Rotary	LTA	Total	OIC/Solo	Training Received	Instructor	Ld.	Dist.
	Dd / Mm		Registration No.	Time		Time									
	__ / __														
	__ / __														
	__ / __														
	__ / __														
	__ / __														
	__ / __														
	__ / __														
	__ / __														
I certify that the entries in this log are true.				Totals this page:											
				Totals brought forward:											
				Totals to date:											

NOTES CONCERNING THE ABOVE FLIGHTS

Flt. No.	Batteries / Fuel	P-f. Ins. C. Cal.	Technical Problems Found / Corrective Action, Status	Remarks and Endorsements

FLT. NO.	DATE Yr ___ Dd / Mm	AIRCRAFT Make/Model Registration No.	FLIGHT LOCATION Takeoff Time	FLIGHT LOCATION Landing Time	FLIGHT DURATION Fixed Wing	FLIGHT DURATION Rotary	FLIGHT DURATION LTA	FLIGHT DURATION Total	FUNCTION OIC/Solo	FUNCTION Training Received	FUNCTION Instructor	DATA Ld.	DATA Dist.
	__/__												
	__/__												
	__/__												
	__/__												
	__/__												
	__/__												
	__/__												
	__/__												

I certify that the entries in this log are true.

Totals this page:										
Totals brought forward:										
Totals to date:										

Flt. No.	Batteries / Fuel	P-f. Ins. C. Cal.	Technical Problems Found / Corrective Action, Status	Remarks and Endorsements

FLT. NO.	DATE	AIRCRAFT	FLIGHT LOCATION		FLIGHT DURATION				FUNCTION			DATA	
	Yr ____	Make/Model	Takeoff	Landing	Fixed Wing	Rotary	LTA	Total	OIC/Solo	Training Received	Instructor	Ld.	Dist.
	Dd / Mm	Registration No.	Time	Time									
	__/__												
	__/__												
	__/__												
	__/__												
	__/__												
	__/__												
	__/__												
	__/__												

I certify that the entries in this log are true.	Totals this page:									
	Totals brought forward:									
	Totals to date:									

Flt. No.	Batteries / Fuel	P-f. Ins. C. Cal.	Technical Problems Found / Corrective Action, Status	Remarks and Endorsements

Flt. No.	Date Yr ___ Dd / Mm	Aircraft Make/Model Registration No.	Flight Location		Flight Duration				Function			Data	
			Takeoff Time	Landing Time	Fixed Wing	Rotary	LTA	Total	OIC/Solo	Training Received	Instructor	Ld.	Dist.
	__/__												
	__/__												
	__/__												
	__/__												
	__/__												
	__/__												
	__/__												
	__/__												
I certify that the entries in this log are true.			Totals this page:										
			Totals brought forward:										
			Totals to date:										

Flt. No.	Batteries / Fuel	P-f. Ins. C. Cal.	Technical Problems Found / Corrective Action, Status	Remarks and Endorsements

FLT. NO.	DATE	AIRCRAFT		FLIGHT LOCATION		FLIGHT DURATION				FUNCTION			DATA	
	Yr ___	Make/Model		Takeoff	Landing	Fixed Wing	Rotary	LTA	Total	OIC/Solo	Training Received	Instructor	Ld.	Dist.
	Dd / Mm	Registration No.		Time	Time									
	__ / __													
	__ / __													
	__ / __													
	__ / __													
	__ / __													
	__ / __													
	__ / __													
	__ / __													
I certify that the entries in this log are true.			Totals this page:											
			Totals brought forward:											
			Totals to date:											

132

Flt. No.	Batteries / Fuel	P-f. Ins. C. Cal.	Technical Problems Found / Corrective Action, Status	Remarks and Endorsements

FLT. NO.	DATE	AIRCRAFT	FLIGHT LOCATION		FLIGHT DURATION				FUNCTION			DATA	
	Yr ___	Make/Model	Takeoff	Landing	Fixed Wing	Rotary	LTA	Total	OIC/Solo	Training Received	Instructor	Ld.	Dist.
	Dd / Mm	Registration No.	Time	Time									
	__/__												
	__/__												
	__/__												
	__/__												
	__/__												
	__/__												
	__/__												
	__/__												

I certify that the entries in this log are true.

Totals this page:	
Totals brought forward:	
Totals to date:	

Flt. No.	Batteries / Fuel	P-f. Ins. C. Cal.	Technical Problems Found / Corrective Action, Status	Remarks and Endorsements

FLT. NO.	DATE Yr ____	AIRCRAFT	FLIGHT LOCATION		FLIGHT DURATION				FUNCTION			DATA	
	Yr ____	Make/Model	Takeoff	Landing	Fixed Wing	Rotary	LTA	Total	OIC/Solo	Training Received	Instructor	Ld.	Dist.
	Dd / Mm	Registration No.	Time	Time									
	__ / __												
	__ / __												
	__ / __												
	__ / __												
	__ / __												
	__ / __												
	__ / __												
	__ / __												
I certify that the entries in this log are true.			Totals this page:										
			Totals brought forward:										
			Totals to date:										

NOTES CONCERNING THE ABOVE FLIGHTS

Flt. No.	Batteries / Fuel	P-f. Ins. C. Cal.	Technical Problems Found / Corrective Action, Status	Remarks and Endorsements

Flt. No.	Date Yr ____ Dd / Mm	Aircraft Make/Model Registration No.	Flight Location		Flight Duration				Function			Data	
			Takeoff Time	Landing Time	Fixed Wing	Rotary	LTA	Total	OIC/Solo	Training Received	Instructor	Ld.	Dist.
	__/__												
	__/__												
	__/__												
	__/__												
	__/__												
	__/__												
	__/__												
	__/__												
I certify that the entries in this log are true.			**Totals this page:**										
			Totals brought forward:										
			Totals to date:										

Flt. No.	Batteries / Fuel	P-f. Ins. C. Cal.	Technical Problems Found / Corrective Action, Status	Remarks and Endorsements

Flt. No.	Date		Aircraft	Flight Location			Flight Duration				Function			Data	
	Yr ____		Make/Model	Takeoff		Landing	Fixed Wing	Rotary	LTA	Total	OIC/Solo	Training Received	Instructor	Ld.	Dist.
	Dd / Mm		Registration No.	Time		Time									
	__ / __														
	__ / __														
	__ / __														
	__ / __														
	__ / __														
	__ / __														
	__ / __														
	__ / __														
I certify that the entries in this log are true.						Totals this page:									
						Totals brought forward:									
						Totals to date:									

NOTES CONCERNING THE ABOVE FLIGHTS

Flt. No.	Batteries / Fuel	P-f. Ins. C. Cal.	Technical Problems Found / Corrective Action, Status	Remarks and Endorsements

FLT. NO.	DATE Yr ____	AIRCRAFT	FLIGHT LOCATION		FLIGHT DURATION				FUNCTION			DATA	
		Make/Model	Takeoff	Landing	Fixed Wing	Rotary	LTA	Total	OIC/Solo	Training Received	Instructor	Ld.	Dist.
	Dd / Mm	Registration No.	Time	Time									
	__/__												
	__/__												
	__/__												
	__/__												
	__/__												
	__/__												
	__/__												
	__/__												

I certify that the entries in this log are true.

Totals this page:								
Totals brought forward:								
Totals to date:								

Flt. No.	Batteries / Fuel	P-f. Ins. C. Cal.	Technical Problems Found / Corrective Action, Status	Remarks and Endorsements
NOTES CONCERNING THE ABOVE FLIGHTS				

Flt. No.	Batteries / Fuel	P-f. Ins. C. Cal.	Technical Problems Found / Corrective Action, Status	Remarks and Endorsements

FLT. NO.	DATE	AIRCRAFT	FLIGHT LOCATION		FLIGHT DURATION				FUNCTION			DATA	
	Yr ____	Make/Model	Takeoff	Landing	Fixed Wing	Rotary	LTA	Total	OIC/Solo	Training Received	Instructor	Ld.	Dist.
	Dd / Mm	Registration No.	Time	Time									
	__ / __												
	__ / __												
	__ / __												
	__ / __												
	__ / __												
	__ / __												
	__ / __												
	__ / __												
I certify that the entries in this log are true.			Totals this page:										
			Totals brought forward:										
			Totals to date:										

144

Flt. No.	Batteries / Fuel	P-f. Ins. C. Cal.	Technical Problems Found / Corrective Action, Status	Remarks and Endorsements

FLT. NO.	DATE Yr ___ Dd / Mm	AIRCRAFT Make/Model Registration No.	FLIGHT LOCATION		FLIGHT DURATION				FUNCTION			DATA	
			Takeoff Time	Landing Time	Fixed Wing	Rotary	LTA	Total	OIC/Solo	Training Received	Instructor	Ld.	Dist.
	__/__												
	__/__												
	__/__												
	__/__												
	__/__												
	__/__												
	__/__												
	__/__												

I certify that the entries in this log are true.

Totals this page:

Totals brought forward:

Totals to date:

146

Flt. No.	Batteries / Fuel	P-f. Ins. C. Cal.	Technical Problems Found / Corrective Action, Status	Remarks and Endorsements

NOTES CONCERNING THE ABOVE FLIGHTS

FLT. NO.	DATE	AIRCRAFT	FLIGHT LOCATION		FLIGHT DURATION				FUNCTION			DATA	
	Yr ____	Make/Model	Takeoff	Landing	Fixed Wing	Rotary	LTA	Total	OIC/Solo	Training Received	Instructor	Ld.	Dist.
	Dd / Mm	Registration No.	Time	Time									
	__ / __												
	__ / __												
	__ / __												
	__ / __												
	__ / __												
	__ / __												
	__ / __												
	__ / __												

I certify that the entries in this log are true.

			Totals this page:										
			Totals brought forward:										
			Totals to date:										

NOTES CONCERNING THE ABOVE FLIGHTS

Flt. No.	Batteries / Fuel	P-f. Ins. C. Cal.	Technical Problems Found / Corrective Action, Status	Remarks and Endorsements

FLT. NO.	DATE Yr ____ Dd / Mm	AIRCRAFT Make/Model Registration No.	FLIGHT LOCATION		FLIGHT DURATION				FUNCTION			DATA	
			Takeoff Time	Landing Time	Fixed Wing	Rotary	LTA	Total	OIC/Solo	Training Received	Instructor	Ld.	Dist.
	__/__												
	__/__												
	__/__												
	__/__												
	__/__												
	__/__												
	__/__												
	__/__												

I certify that the entries in this log are true.

Totals this page:	
Totals brought forward:	
Totals to date:	

NOTES CONCERNING THE ABOVE FLIGHTS

Flt. No.	Batteries / Fuel	P-f. Ins. C. Cal.	Technical Problems Found / Corrective Action, Status	Remarks and Endorsements

Flt. No.	Date Yr ____ Dd / Mm	Aircraft Make/Model Registration No.	Flight Location		Flight Duration				Function			Data	
			Takeoff Time	Landing Time	Fixed Wing	Rotary	LTA	Total	OIC/Solo	Training Received	Instructor	Ld.	Dist.
	__/__												
	__/__												
	__/__												
	__/__												
	__/__												
	__/__												
	__/__												
	__/__												
I certify that the entries in this log are true.				Totals this page:									
				Totals brought forward:									
				Totals to date:									

Flt. No.	Batteries / Fuel	P-f. Ins. C. Cal.	Technical Problems Found / Corrective Action, Status	Remarks and Endorsements

FLT. NO.	DATE Yr ____ Dd / Mm	AIRCRAFT Make/Model Registration No.	FLIGHT LOCATION Takeoff Time	Landing Time	FLIGHT DURATION Fixed Wing	Rotary	LTA	Total	FUNCTION OIC/Solo	Training Received	Instructor	DATA Ld.	Dist.
	__/__												
	__/__												
	__/__												
	__/__												
	__/__												
	__/__												
	__/__												
	__/__												

I certify that the entries in this log are true.

Totals this page:

Totals brought forward:

Totals to date:

NOTES CONCERNING THE ABOVE FLIGHTS

Flt. No.	Batteries / Fuel	P-f. Ins. C. Cal.	Technical Problems Found / Corrective Action, Status	Remarks and Endorsements

Flt. No.	Date Yr ____ Dd / Mm	Aircraft Make/Model Registration No.	Flight Location		Flight Duration				Function			Data	
			Takeoff Time	Landing Time	Fixed Wing	Rotary	LTA	Total	OIC/Solo	Training Received	Instructor	Ld.	Dist.
	__ / __												
	__ / __												
	__ / __												
	__ / __												
	__ / __												
	__ / __												
	__ / __												
	__ / __												
I certify that the entries in this log are true.			Totals this page:										
			Totals brought forward:										
			Totals to date:										

Flt. No.	Batteries / Fuel	P-f. Ins. C. Cal.	Technical Problems Found / Corrective Action, Status	Remarks and Endorsements

FLT. NO.	DATE Yr ___ Dd / Mm	AIRCRAFT Make/Model Registration No.	FLIGHT LOCATION		FLIGHT DURATION				FUNCTION			DATA	
			Takeoff Time	Landing Time	Fixed Wing	Rotary	LTA	Total	OIC/Solo	Training Received	Instructor	Ld.	Dist.
	__/__												
	__/__												
	__/__												
	__/__												
	__/__												
	__/__												
	__/__												
	__/__												

I certify that the entries in this log are true.

Totals this page:			
Totals brought forward:			
Totals to date:			

NOTES CONCERNING THE ABOVE FLIGHTS

Flt. No.	Batteries / Fuel	P-f. Ins. C. Cal.	Technical Problems Found / Corrective Action, Status	Remarks and Endorsements

FLT. NO.	DATE	AIRCRAFT	FLIGHT LOCATION		FLIGHT DURATION				FUNCTION			DATA	
	Yr ____	Make/Model	Takeoff	Landing	Fixed Wing	Rotary	LTA	Total	OIC/Solo	Training Received	Instructor	Ld.	Dist.
	Dd / Mm	Registration No.	Time	Time									
	__/__												
	__/__												
	__/__												
	__/__												
	__/__												
	__/__												
	__/__												
	__/__												
I certify that the entries in this log are true.			Totals this page:										
			Totals brought forward:										
			Totals to date:										

NOTES CONCERNING THE ABOVE FLIGHTS

Flt. No.	Batteries / Fuel	P-f. Ins. C. Cal.	Technical Problems Found / Corrective Action, Status	Remarks and Endorsements

FLT. NO.	DATE Yr ___ Dd / Mm	AIRCRAFT Make/Model Registration No.	FLIGHT LOCATION Takeoff — Time	Landing — Time	FLIGHT DURATION Fixed Wing	Rotary	LTA	Total	FUNCTION OIC/Solo	Training Received	Instructor	DATA Ld.	Dist.
	__ / __												
	__ / __												
	__ / __												
	__ / __												
	__ / __												
	__ / __												
	__ / __												
	__ / __												
I certify that the entries in this log are true.			Totals this page:										
			Totals brought forward:										
			Totals to date:										

NOTES CONCERNING THE ABOVE FLIGHTS

Flt. No.	Batteries / Fuel	P-f. Ins. C. Cal.	Technical Problems Found / Corrective Action, Status	Remarks and Endorsements

FLT. NO.	DATE	AIRCRAFT	FLIGHT LOCATION		FLIGHT DURATION				FUNCTION			DATA	
	Yr ____	Make/Model	Takeoff	Landing	Fixed Wing	Rotary	LTA	Total	OIC/Solo	Training Received	Instructor	Ld.	Dist.
	Dd / Mm	Registration No.	Time	Time									
	__ / __												
	__ / __												
	__ / __												
	__ / __												
	__ / __												
	__ / __												
	__ / __												
	__ / __												

I certify that the entries in this log are true.

Totals this page:		
Totals brought forward:		
Totals to date:		

NOTES CONCERNING THE ABOVE FLIGHTS

Flt. No.	Batteries / Fuel	P-f. Ins. C. Cal.	Technical Problems Found / Corrective Action, Status	Remarks and Endorsements

FLT. NO.	DATE Yr ___ Dd / Mm	AIRCRAFT Make/Model Registration No.	FLIGHT LOCATION		FLIGHT DURATION				FUNCTION			DATA	
			Takeoff Time	Landing Time	Fixed Wing	Rotary	LTA	Total	OIC/Solo	Training Received	Instructor	Ld.	Dist.
	__/__												
	__/__												
	__/__												
	__/__												
	__/__												
	__/__												
	__/__												
	__/__												

I certify that the entries in this log are true.

Totals this page:			
Totals brought forward:			
Totals to date:			

Flt. No.	Batteries / Fuel	P-f. Ins. C. Cal.	Technical Problems Found / Corrective Action, Status	Remarks and Endorsements

Flt. No.	Date Yr ___ Dd / Mm	Aircraft Make/Model Registration No.	Flight Location		Flight Duration				Function			Data	
			Takeoff Time	Landing Time	Fixed Wing	Rotary	LTA	Total	OIC/Solo	Training Received	Instructor	Ld.	Dist.
	__/__												
	__/__												
	__/__												
	__/__												
	__/__												
	__/__												
	__/__												
	__/__												

I certify that the entries in this log are true.

Totals this page:	
Totals brought forward:	
Totals to date:	

Flt. No.	Batteries / Fuel	P-f. Ins. C. Cal.	Technical Problems Found / Corrective Action, Status	Remarks and Endorsements

FLT. NO.	DATE Yr ____ Dd / Mm	AIRCRAFT Make/Model Registration No.	FLIGHT LOCATION Takeoff Time	Landing Time	FLIGHT DURATION Fixed Wing	Rotary	LTA	Total	FUNCTION OIC/Solo	Training Received	Instructor	DATA Ld.	Dist.
	__/__												
	__/__												
	__/__												
	__/__												
	__/__												
	__/__												
	__/__												

I certify that the entries in this log are true.

Totals this page:							
Totals brought forward:							
Totals to date:							

NOTES CONCERNING THE ABOVE FLIGHTS

Flt. No.	Batteries / Fuel	P-f. Ins. C. Cal.	Technical Problems Found / Corrective Action, Status	Remarks and Endorsements

FLT. No.	DATE Yr ____	AIRCRAFT	FLIGHT LOCATION		FLIGHT DURATION				FUNCTION			DATA	
		Make/Model	Takeoff	Landing	Fixed Wing	Rotary	LTA	Total	OIC/Solo	Training Received	Instructor	Ld.	Dist.
	Dd / Mm	Registration No.	Time	Time									
	__ / __												
	__ / __												
	__ / __												
	__ / __												
	__ / __												
	__ / __												
	__ / __												
I certify that the entries in this log are true.			Totals this page:										
			Totals brought forward:										
			Totals to date:										

172

NOTES CONCERNING THE ABOVE FLIGHTS

Flt. No.	Batteries / Fuel	P-f. Ins. C. Cal.	Technical Problems Found / Corrective Action, Status	Remarks and Endorsements

Flt. No.	Date Yr ____ Dd / Mm	Aircraft Make/Model Registration No.	Flight Location		Flight Duration				Function			Data	
			Takeoff Time	Landing Time	Fixed Wing	Rotary	LTA	Total	OIC/Solo	Training Received	Instructor	Ld.	Dist.
	__/__												
	__/__												
	__/__												
	__/__												
	__/__												
	__/__												
	__/__												
	__/__												
I certify that the entries in this log are true.			Totals this page:										
			Totals brought forward:										
			Totals to date:										

NOTES CONCERNING THE ABOVE FLIGHTS

Flt. No.	Batteries / Fuel	P-f. Ins. C. Cal.	Technical Problems Found / Corrective Action, Status	Remarks and Endorsements

FLT. NO.	DATE		AIRCRAFT	FLIGHT LOCATION			FLIGHT DURATION				FUNCTION			DATA	
	Yr ____		Make/Model	Takeoff	Landing		Fixed Wing	Rotary	LTA	Total	OIC/Solo	Training Received	Instructor	Ld.	Dist.
	Dd / Mm		Registration No.	Time	Time										
	__/__														
	__/__														
	__/__														
	__/__														
	__/__														
	__/__														
	__/__														
	__/__														

I certify that the entries in this log are true.	Totals this page:									
	Totals brought forward:									
	Totals to date:									

NOTES CONCERNING THE ABOVE FLIGHTS

Flt. No.	Batteries / Fuel	P-f. Ins. C. Cal.	Technical Problems Found / Corrective Action, Status	Remarks and Endorsements

FLT. NO.	DATE		AIRCRAFT	FLIGHT LOCATION			FLIGHT DURATION				FUNCTION			DATA	
	Yr ____		Make/Model	Takeoff	Landing		Fixed Wing	Rotary	LTA	Total	OIC/Solo	Training Received	Instructor	Ld.	Dist.
	Dd / Mm		Registration No.	Time	Time										
	__/__														
	__/__														
	__/__														
	__/__														
	__/__														
	__/__														
	__/__														
	__/__														

I certify that the entries in this log are true.	Totals this page:							
	Totals brought forward:							
	Totals to date:							

NOTES CONCERNING THE ABOVE FLIGHTS

Flt. No.	Batteries / Fuel	P-f. Ins. C. Cal.	Technical Problems Found / Corrective Action, Status	Remarks and Endorsements

FLT. NO.	DATE	AIRCRAFT	FLIGHT LOCATION		FLIGHT DURATION				FUNCTION			DATA	
	Yr ____	Make/Model	Takeoff	Landing	Fixed Wing	Rotary	LTA	Total	OIC/Solo	Training Received	Instructor	Ld.	Dist.
	Dd / Mm	Registration No.	Time	Time									
	__/__												
	__/__												
	__/__												
	__/__												
	__/__												
	__/__												
	__/__												
	__/__												

I certify that the entries in this log are true.

Totals this page:

Totals brought forward:

Totals to date:

180

Flt. No.	Batteries / Fuel	P-f. Ins. C. Cal.	Technical Problems Found / Corrective Action, Status	Remarks and Endorsements

FLT. NO.	DATE		AIRCRAFT	FLIGHT LOCATION			FLIGHT DURATION				FUNCTION			DATA	
	Yr ____		Make/Model	Takeoff	Landing		Fixed Wing	Rotary	LTA	Total	OIC/Solo	Training Received	Instructor	Ld.	Dist.
	Dd / Mm		Registration No.	Time	Time										
	__/__														
	__/__														
	__/__														
	__/__														
	__/__														
	__/__														
	__/__														
	__/__														
I certify that the entries in this log are true.					Totals this page:										
					Totals brought forward:										
					Totals to date:										

NOTES CONCERNING THE ABOVE FLIGHTS

Flt. No.	Batteries / Fuel	P-f. Ins. C. Cal.	Technical Problems Found / Corrective Action, Status	Remarks and Endorsements

PER-AIRCRAFT TOTALS

(requires manual review and summing of logged data for each aircraft flown)

	AIRCRAFT	FLIGHT TIME				FUNCTION			DATA
	Registration No.	Fixed	Rotary	LTA	Total	OIC/Solo	Training	Instructor	Ld.
1									
2									
3									
4									
5									
6									
7									
8									
9									
10									
11									
12									
Totals should equal those on pg 182:									

PART 3: RECORD OF PILOT TRAINING

FLIGHT AND GROUND TRAINING RECORD

Date	Course	Duration	Test Score	Pass Y / N	Instructor / Examiner Name, Signature, Certificate Number

FLIGHT AND GROUND TRAINING RECORD, CONTINUED

Date	Course	Duration	Test Score	Pass Y / N	Instructor / Examiner Name, Signature, Certificate Number

TRAINING CERTIFICATION

I certify that _____
has satisfactorily completed the aviation training course/
program titled

_____

Signed _____ Date _____

Cert. # _____ Expiration _____

I certify that _____
has satisfactorily completed the aviation training course/
program titled

_____

Signed _____ Date _____

Cert. # _____ Expiration _____

I certify that _____
has satisfactorily completed the aviation training course/
program titled

_____

Signed _____ Date _____

Cert. # _____ Expiration _____

I certify that _____
has satisfactorily completed the aviation training course/
program titled

_____

Signed _____ Date _____

Cert. # _____ Expiration _____

TRAINING CERTIFICATION, CONTINUED

I certify that _____
has satisfactorily completed the aviation training course/
program titled

_____

Signed _____ Date _____

Cert. # _____ Expiration _____

I certify that _____
has satisfactorily completed the aviation training course/
program titled

_____

Signed _____ Date _____

Cert. # _____ Expiration _____

I certify that _____
has satisfactorily completed the aviation training course/
program titled

_____

Signed _____ Date _____

Cert. # _____ Expiration _____

I certify that _____
has satisfactorily completed the aviation training course/
program titled

_____

Signed _____ Date _____

Cert. # _____ Expiration _____

I certify that _____
has satisfactorily completed the aviation training course/
program titled

_____

Signed _____ Date _____

Cert. # _____ Expiration _____

I certify that _____
has satisfactorily completed the aviation training course/
program titled

_____

Signed _____ Date _____

Cert. # _____ Expiration _____

I certify that _____
has satisfactorily completed the aviation training course/
program titled

_____

Signed _____ Date _____

Cert. # _____ Expiration _____

I certify that _____
has satisfactorily completed the aviation training course/
program titled

_____

Signed _____ Date _____

Cert. # _____ Expiration _____

ADDITIONAL NOTES

Printed in September 2021
by Rotomail Italia S.p.A., Vignate (MI) - Italy